2nd Edition. Copyright © 2016 by David R. Dixon / Illustrated by Phil Jones / Edited by Mark Brannan

To contact the author, please visit the Goodnight Marines Facebook page at **https://www.facebook.com/goodnightmarines/**

All rights reserved. No part of this publication may be reproduced, photocopied, stored in a retrieval system, or transmitted in any form or by any means - for example, electronic, photocopy, audio recording - without the prior written permission of the publisher, Callsign Enterprises, except by a reviewer who may quote a few lines or use one illustration in a review.

ISBN-13: 978-1-941698-04-4   **ISBN-10: 1-941698-04-2**

Printed in Dallas. Don't Mess With Texas!

To contact the publisher, please send an email to **info@callsignenterprises.com**

Callsign Enterprises is a certified Historically Underutilized Business by the State of Texas and a certified Veteran Owned Small Business by the United States Department of Veterans Affairs.

A portion of the proceeds from this book will benefit Tillman Scholars through the Pat Tillman Foundation. Please visit **www.pattillmanfoundation.org**

Neither the United States Marine Corps nor any other component of the Department of Defense has approved, endorsed, or authorized this book.

To Jaclyn,

"I have found the one my soul loves"

- Song of Solomon 3:4

In a land overseas, in a far away place...

There was a brave Marine, guarding his base.

And behind him stood a silent platoon.

And there were planes flying over the Moon

And there were two shiny stars, and two silver bars.

And a pair of boots, and two yellow feet.
And a pair of Dress Blues, and a mighty Navy fleet.

And a flag in black volcanic sand, and the President's Own Band.

And the finest dog in all the land

Now I lay in my rack, with my sleepy head down,
And say Goodnight to my Marines who keep me safe and sound.

Goodnight Smedley, John, and Dan. Goodnight Archie, the Grand Ole Man.

Goodnight to the guards of Heaven's streets,
Goodnight Sergeant Major who never retreats.

Goodnight recruits on Parris Island afar.
Goodnight Chesty Puller, wherever you are.

Goodnight Eagle, Globe, and Anchor. Goodnight Captain, Goodnight tanker.

Goodnight Tun Tavern and Semper Fi.

Goodnight trucks, Goodnight supply.

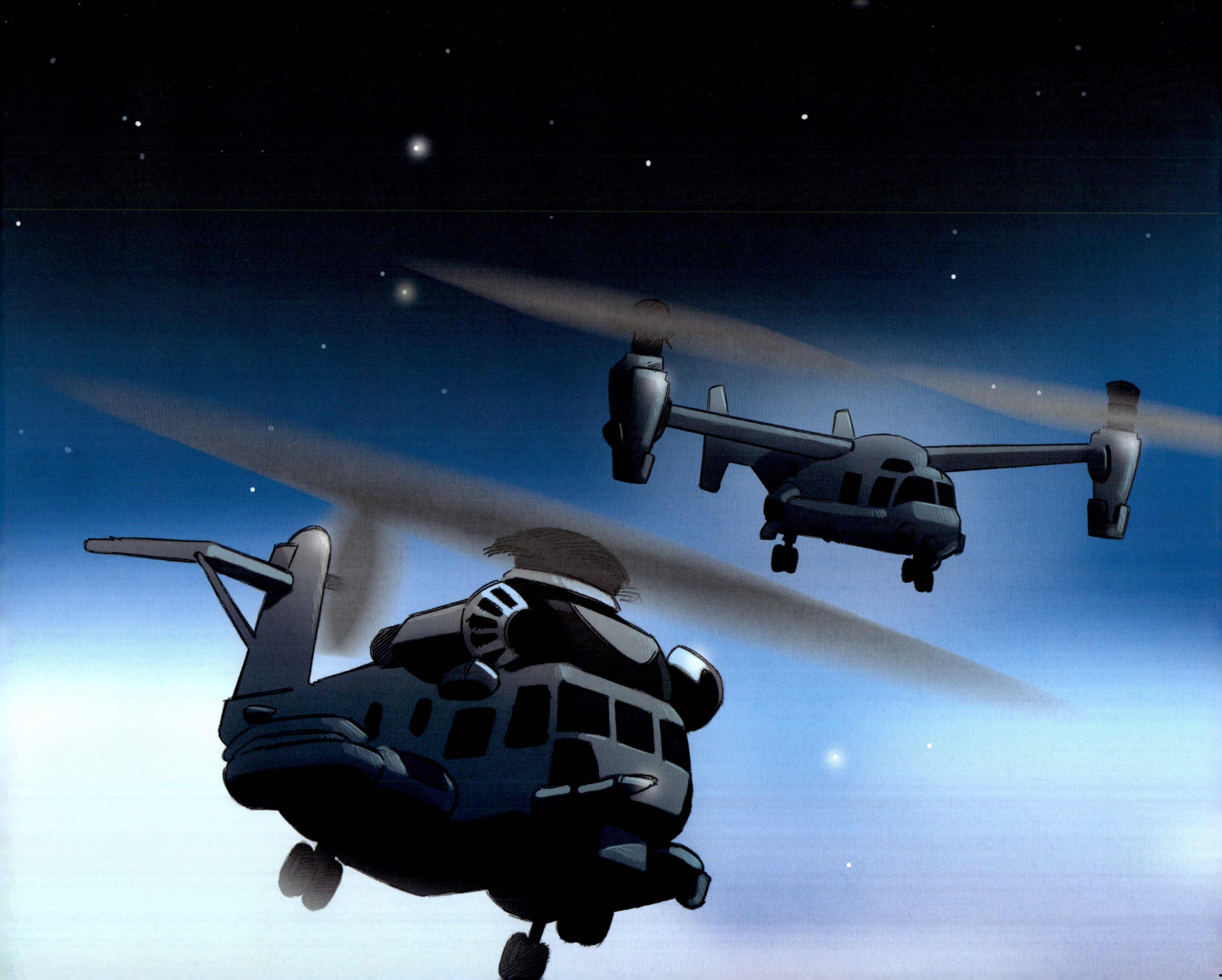

Goodnight artillery with your boom so loud,
Goodnight helicopters, floating on a cloud.

Goodnight boots, Goodnight flags, Goodnight radios, Goodnight dog tags

Goodnight band, softly playing Taps, Goodnight compass, Goodnight maps.

Goodnight Blues with scarlet thread,
Goodnight Tuffy Hound on the end of my bed.

Goodnight Tripoli, Belleau Wood, and Saipan.
Goodnight to my Dad in Afghanistan.

Goodnight heroes who answer the call.
Goodnight Marines, and those that gave all.

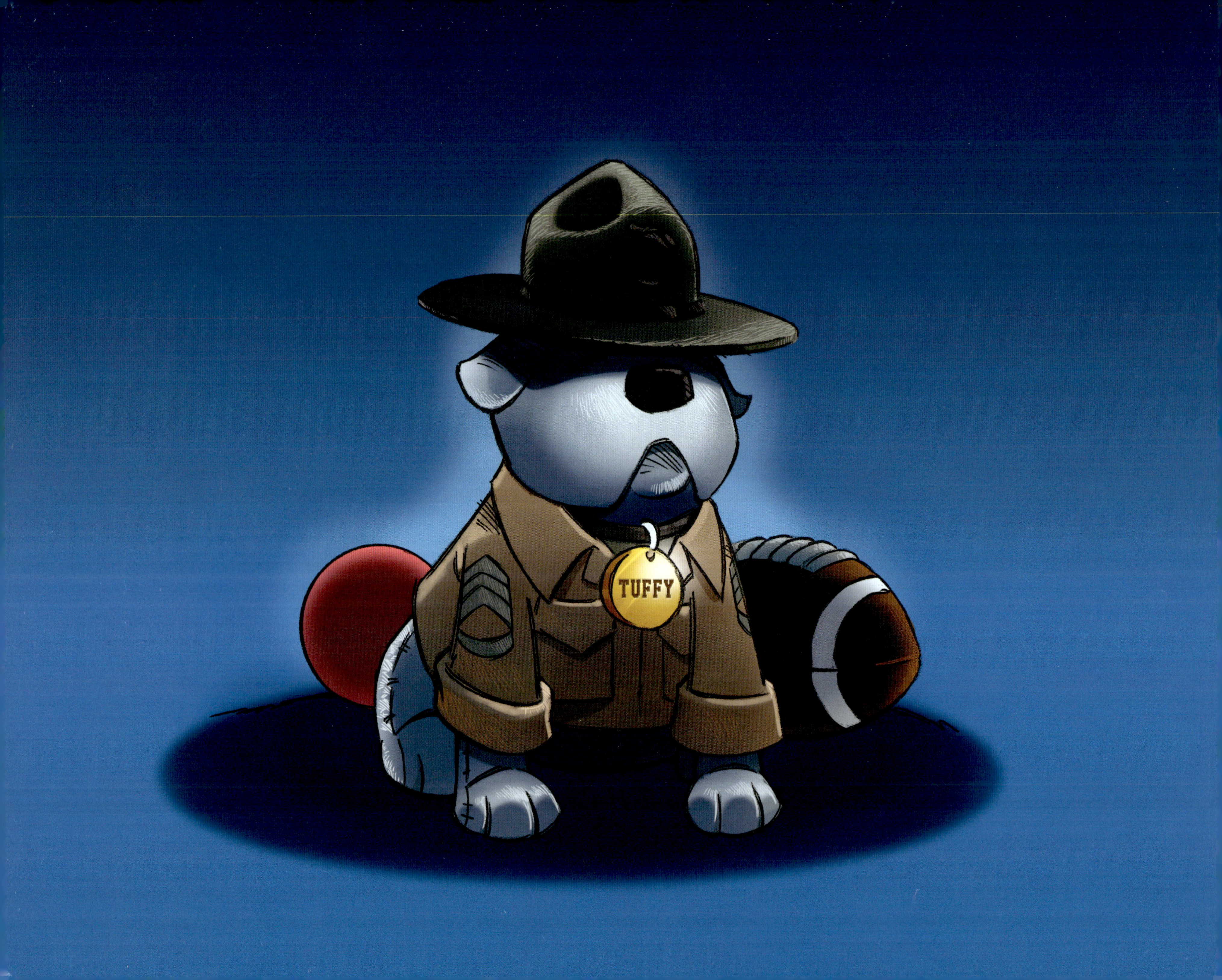
TUFFY

The Battle

The Base

The Marines' Hymn

From the Halls of Montezuma
To the shores of Tripoli;
We fight our country's battles
In the air, on land, and sea;
First to fight for right and freedom
And to keep our honor clean;
We are proud to claim the title
Of United States Marine.

Our flag's unfurled to every breeze
From dawn to setting sun;
We have fought in every clime and place
Where we could take a gun;
In the snow of far off Northern lands
And in sunny tropic scenes,
You will find us always on the job
The United States Marines.

Here's health to you and to our Corps
Which we are proud to serve;
In many a strife we've fought for life
And never lost our nerve.
If the Army and the Navy
Ever look on Heaven's scenes,
They will find the streets are guarded
By United States Marines.

About *Goodnight Marines*

As a veteran I feel honored to represent the Marine Corps and act as an ambassador for the traditions that I hold so dear. I believe that Marines, both in and out of uniform, must engage the younger generation and share our proud culture of leadership. In this spirit I authored *Goodnight Marines*. The inspiration for many of these verses and illustrations came during the difficult times away from my family during two combat deployments to Iraq.

Civilians who read *Goodnight Marines* will certainly appreciate the patriotism, heartfelt emotions, and captivating illustrations by Phil Jones (Desert Storm veteran and former Disney artist). Leathernecks sharing this with young children will quickly realize the depths of symbolism. In this book a young child gazes proudly at a photograph of his father - currently deployed overseas to Afghanistan. The father stands watch over his combat outpost underneath the very same moon (with an Eagle, Globe, and Anchor) that is shining on his son. The child's room is adorned with toy tanks, planes, radios, and trucks that he plays with to feel connected to his father. Then, as the child says goodnight to his dad and all of his military memorabilia, the toys come alive in his vibrant imagination.

While canvasing the child's room, *Goodnight Marines* references many Marine Corps occupational specialties, including Infantry, Aviation, Communications, Supply, Logistics, Artillery, the Band, and even Toys for Tots. Thus, the book offers a chance for all Marines to show their child "This is what mommy / daddy does at work." This book also intentionally incorporates a diversity of ranks - junior enlisted, NCO, SNCO, Sergeant Major, Warrant Officer, and Officer - so that all Marines may find personal examples to feel proud of when reading to their little loved ones. In fact this poem leaves the specific rank and occupation of the father unmentioned, symbolizing that every Marine is a rifleman. *Goodnight Marines* also remembers heroes such as Dan Daly and famous battles like Iwo Jima and Saipan.

The kids, of course, love the quiet protagonist - the child's best friend and stuffed animal, Teufel Hunden (Devil Dog). Unfortunately, this young boy cannot yet pronounce such a difficult German word, so he has named his friend "Tuffy Hound." Tuffy was given to the child by his dad before deploying to Afghanistan. Just like the father guards his combat outpost, Tuffy provides the child with warmth and security, watching over the room as the child sleeps. Tuffy is a Staff Sergeant and wears the Smokey Bear cover of a Marine Drill Instructor, symbolizing the mentorship, guidance, and (albeit sometimes tough) love that DI's show as they become parental figures to recruits during Boot Camp. The real spirit of *Goodnight Marines* culminates on the final page, where the moon shines on the father's picture next to his son - sleeping safely in his room with Tuffy Hound. I encourage the audience to conclude their nightly reading by softly singing the Marines' Hymn. My vision is that before deployments, Marines will record themselves reading this book out loud, and then their spouse can play the video while mommy / daddy is gone in order for the kids to stay connected and establish a routine. I hope that in some small way this helps bring Marine families closer together and that my fellow warriors will treasure reading *Goodnight Marines* to their children as much as my wife and I do.

Semper Fidelis,
David R. Dixon

Major David Dixon is a Marine Corps AH-1W Cobra helicopter pilot and an Iraq War Veteran. David graduated with honors from Texas A&M University and earned his Master's Degrees from the Harvard Graduate School of Education and the Stanford Graduate School of Business. His book, *Call in the Air*, won the Robert A. Gannon award for distinguished poetry from the Marine Corps Heritage Foundation. David is married to the love of his life, Jackie Dixon, and they currently reside in New Orleans, Louisiana.

Phil Jones is a graduate of the Ringling College of Art and Design. He has been working as an artist for over 30 years including a position at Disney Feature Animation. Mr. Jones is also a Desert Storm Veteran serving with the 160th Special Operations Aviation Regiment (ABN). He now resides in Sarasota, Florida with his wife Norayda and their son, Austin.

Goodnight Marines Glossary

- **Silent Platoon** - The world-renowned Silent Drill Team is a 24-man rifle platoon that puts on a captivating performance without vocal commands. In 1948 they gave what was supposed to be a one-time performance at the Marine Barracks' parades, but they received such a positive reaction that they later became a regular part of the show. Now they perform at the Evening Parade during the summer every Friday evening from May through August. This parade features not only the Silent Drill Team, but also "The President's Own" United States Marine Band and "The Commandant's Own" the United States Marine Drum and Bugle Corps. The Marine Barracks are located at 8th and I near the Capitol and the Washington Navy Yard. President Thomas Jefferson chose this location (formerly Square 927) in 1801. In 1806 construction for the Home of the Commandants was completed - it is the oldest public building in continuous use in Washington, D.C. Legend holds that when the British burned D.C. on August 24, 1814 (during the War of 1812), they intentionally spared the home and the 8th and I Barracks out of reverence for the Marines' valor during the Battle of Bladensburg, Maryland earlier that day. However, President James Madison and many others from the federal government had been present at Bladensburg and were nearly captured - that battle is often referred to as "The greatest disgrace ever dealt to American arms."

- **Two Shiny Stars** - The Bronze Star Medal: Awarded for acts of meritorious achievement, meritorious service, or heroism in combat. The Silver Star Medal: The third-highest military award for valor, given for gallantry in combat.

- **Two Silver Bars** - The Secretary of the Navy approves a warrant for a staff non-commissioned officer to be appointed a warrant officer, who typically serves as a technical advisor (although some eventually become commissioned officers called "Limited Duty Officers"). When a Chief Warrant Officer serves as an infantry weapons officer, he carries the title "Marine Gunner," and he wears a bursting bomb insignia on his left collar. The bars in this book are of a CWO-5 (the highest rank).

- **Two Yellow Feet** - Upon arrival to San Diego or Parris Island, civilian recruits are ordered off the bus to stand at attention on yellow footprints painted on the asphalt. The Drill Instructors tell the recruits, "You have just taken the first step toward becoming a member of the world's finest fighting force." On these iconic footprints, the transformation begins.

- **A Pair of Dress Blues (with Scarlet Thread)** - The Marine Dress Blue uniform consists of a long-sleeved midnight blue coat (enlisted members have red trim) - comparable in use to civilian black tie. Corporals and above wear a scarlet "blood stripe" down the trouser leg. Legend holds that the blood stripe was added after the 1847 Battle of Chapultepec, where Marines sustained a high casualty rate. This battle also inspired the opening of the Marines' Hymn, "From the Halls of Montezuma."

- **Mighty Navy Fleet** - The Marine Corps belongs to the Department of the Navy, and as "Soldiers of the Sea," Marines are inextricably linked to their naval heritage. In the past, Marines wore thick leather collars for protection against slashes as they boarded enemy vessels - thus earning the nickname "Leathernecks." Sailors have also jocularly referred to Marines as

Goodnight Marines Glossary

"Gyrenes" (probably a combination of the words GI and Marine) and "Jarheads" (likely from the resemblance of the high collar of the Marine Dress Blue uniforms to Mason jars - which were sometimes made from a blue-colored glass).

- **Flag in Black Volcanic Sand** - Near the end of World War II, Marines raised the American flag on Mount Suribachi during the Battle of Iwo Jima. Photographer Joe Rosenthal, a war correspondent with the Associated Press, captured the iconic moment and eventually won the Pulitzer Prize for his photo. The picture was later used to sculpt the Marine Corps War Memorial, which sits adjacent to Arlington National Cemetery on the banks of the Potomac River near Washington, D.C.

- **President's Own Band** - President John Adams signed an Act of Congress in 1798 establishing the United States Marine Band, at the time consisting of "32 drummers and fifers." They performed during the 1801 inauguration of Thomas Jefferson - a music enthusiast credited with coining the title "The President's Own" – and have played during every presidential inauguration since. Boasting famous former members such as John Philip Sousa, they are the oldest continuously performing professional musical organization in the U.S. The President's Own (now about 130 members) do not attend Boot Camp, unlike their peers in the Marine Drum & Bugle Corps and Fleet Marine Bands, who go through traditional Recruit Training.

- **The Finest Dog in All the Land** - According to legend, German soldiers used the moniker Teufel Hunden (Devil Dogs) to describe U.S. Marines who fought with ferocity during the Battle of Belleau Wood in 1918. Accordingly, Brigadier General Smedley Butler adopted the purebred English bulldog as the Corps' official mascot in 1922 - originally called King Bulwark but then renamed Jiggs. From the 1930s to the 1950s, the bulldog mascots were named Smedley, but since 1957 the mascots have been called Chesty. You can read more about this book's mascot, Tuffy Hound, in the "About" section.

- **Smedley** - Major General Smedley D. Butler (known as "Ol' Gimlet Eye" for his battle stare) was born July 30, 1881. His father was a Congressman from Pennsylvania and Chairman of the House Naval Affairs Committee. He became a second lieutenant as a teenager during the Spanish-American War. Smedley Butler received the Medal of Honor for his heroism in Vera Cruz (1914) and again for his gallantry in hand-to-hand combat with the Caco resistance at Fort Riviere, Haiti (1915).

- **John** - This refers to both Lieutenant General John Archer Lejeune and Gunnery Sergeant John Basilone. John A. Lejeune was born in Pointe Coupee, Louisiana on January 10, 1867, and served as the 13th Commandant of the Marine Corps. During WWI he was the first Marine to command an Army division in war, and after the Armistice he led his division during the march into Germany. He was highly decorated by the French government for his leadership, earning the Legion of Honor and the Croix de Guerre. He died in 1942 and is buried in Arlington National Cemetery. Many Marines refer to Lieutenant General Lejeune as "the Greatest of all Leathernecks." Gunnery Sergeant John Basilone was born in Buffalo, New York on November 4, 1916. He enlisted in the Army when he turned 18, and after his three year enlistment he worked as a truck driver in Maryland before enlisting in the Marine Corps in July 1940. He received the Medal of Honor for heroism at the

Goodnight Marines Glossary

Battle of Guadalcanal. Gunnery Sergeant Basilone died in combat on February 19, 1945, during the first day of the Battle of Iwo Jima. He was posthumously awarded the Navy Cross for his extraordinary heroism.

- **Dan** - Sergeant Major Daniel Joseph "Dan" Daly (1873 – 1937) is one of only 19 men (along with Smedley Butler) to receive the Medal of Honor twice - for his valor in the Battle of Peking during the Boxer Rebellion (China, 1900) and then during the Battle of Fort Dipitie, Haiti (1915). He also earned the Navy Cross for his heroism during the Battle of Belleau Wood.

- **The Grand Old Man** - Colonel (Brevet Brigadier General) Archibald Henderson (1783 - 1859), known as the "Grand Old Man of the Marine Corps," served for over 54 years and was the longest serving Commandant (the highest ranking General in the Marine Corps). According to lore, he lived in the Home of the Commandants so long (over 38 years) that he forgot it was actually owned by the government and attempted to will the home to his heirs.

- **Guards of Heaven's Streets** - This is a reference to a line in the Marines' Hymn (the oldest official song in the United States military), and this picture fittingly shows the Toys for Tots campaign. Founded in 1947 by Major Bill Hendricks, Toys for Tots is run jointly by the Marine Corps Reserve and the Marine Corps League and donates toys during the Christmas season.

- **Sergeant Major Who Never Retreats** - Archibald Sommers was the first Marine to be appointed to the grade of Sergeant Major in 1801 (at that time it was a solitary post). The rank was made permanent for the Marine Corps in 1833, and five Marines held the rank of Sergeant Major by 1899. The rank was abolished in 1946, but then re-introduced in 1954. Wilbur Bestwick was appointed in 1957 to serve in the post of Sergeant Major of the Marine Corps - the first such position in the United States military. The SMMC serves as the senior enlisted advisor to the Commandant, and the rank has its own distinctive insignia - featuring the Eagle, Globe, and Anchor alongside two stars.

- **Parris Island Afar** - Enlisted recruit training occurs in San Diego, CA and Parris Island, SC. No matter the distance from their hometown, recruits often feel they are in a faraway land while enduring the grueling transformation to become a Marine.

- **Chesty** - Lieutenant General Lewis Burwell "Chesty" Puller (1898 – 1971) was one of the most decorated Marines in history, and the only Marine to receive five Navy Crosses (2nd highest award). He saw combat in WWII, Korea, Haiti, and Nicaragua. Marines have been known to recite the phrase "Goodnight Chesty Puller, wherever you are" before hitting the rack, and you will often hear Marines during their physical training motivating each other by yelling "One more for Chesty!"

- **Eagle, Globe, and Anchor** - Brigadier General Jacob Zeilin, 7th Commandant of the Marine Corps, approved a version of the EGA as the Corps' official emblem in 1868. After several cosmetic variations, President Eisenhower signed Executive Order 10538 in 1954, establishing the current EGA insignia on a scarlet background encircled by a gold rope. The Marine Corps has different insignia for officers and enlisted. Gold and silver colored metal make up the officer's dress insignia while

Goodnight Marines Glossary

the enlisted insignia is comprised of a gold colored metal and shows Cuba (which the officer insignia does not). Many legends exist for why this is so (such as the composition of Marine forces during the Spanish-American War). However the real reason is more pragmatic. Enlisted insignia is stamped from a single piece of metal, while the officer insignia is composed of several different metals, and mounting the tiny island of Cuba was too cumbersome. In this book the Enlisted EGA appears at the beginning and the Officer EGA is displayed at the end, because by Marine Corps leadership doctrine, officers should always put the needs of their enlisted Marines before their own. In other words, "Officers eat last."

- **Goodnight Captain** - This picture features the Leftwich Trophy, given each year to the most outstanding Marine Captain serving with the ground forces of the Fleet Marine Force. The award honors the memory of Lieutenant Colonel William Groom Leftwich Jr., who served as the Commanding Officer of 1st Reconnaissance Battalion, 1st Marine Division (Reinforced) during Vietnam. He died in a helicopter crash during a combat mission in November 1970. A partial list of his awards includes the Navy Cross, the Silver Star (posthumous), the Legion of Merit with Combat "V," and the Purple Heart.

- **Tun Tavern** - On November 10, 1775, the Second Continental Congress commissioned Samuel Nicholas to raise two battalions of Marines in order to augment naval forces in the Revolutionary War. Captain Nicholas visited many inns and taverns throughout Philadelphia during his recruiting drive, and one of his first recruits was Robert Mullan, the owner of Tun Tavern, which legend holds as the birthplace of the Corps. After much success in battle, the Marine Corps was abolished after the Revolutionary War for economic reasons. Then, on July 11, 1798, President John Adams approved a Congressional bill that recreated the Marine Corps under the Secretary of the Navy. From 1799-1921, Marines celebrated their birthday (Marine Corps Day) on July 11th. In 1921, the 13th Commandant, Major General John A. Lejeune, issued Marine Corps Order No. 47, Series 1921 - permanently recognizing the Marine Corps' birthday as November 10, 1775.

- **Semper Fi** - In 1883, Colonel Charles McCawley, the 8th Commandant, adopted "Semper Fidelis" (Always Faithful) as the Marines' official motto. Before this, the Corps had used three mottos, all colloquial rather than formal. The first, "Fortitudine" ("With Fortitude", or "With Courage") predated the War of 1812. The next motto, "By Sea and By Land," was likely based on the British Royal Marines' motto "Per Mare, Per Terram." The Corps' third motto was "To the Shores of Tripoli," which honored First Lieutenant Presley O'Bannon's raising of the American flag over the walled city of Derne, Tripoli in 1805. This was the first land battle for the U.S. on foreign soil, and thus the first time the American flag was raised over a conquered foreign city. In 1848 the motto was updated to "From the Halls of Montezuma, to the Shores of Tripoli" – which subsequently became the opening line in the Marines' Hymn. Since 1883, the Semper Fidelis motto has encouraged Marines to remain faithful to the United States, the Corps, the mission at hand, and to their brothers and sisters in arms.

Goodnight Marines Glossary

- **Tripoli** - The Navy was ordered by President Jefferson to the Mediterranean in 1801 to disrupt the frequent raids against U.S. ships by pirates from Algeria, Morocco, Tripolitania, and Tunis (The Barbary States). The pirates would capture American sailors and cargo and then ransom it back at extortionate prices. William Eaton, a former Army officer and the "Navy Agent for the several Barbary Regencies," led a force of 7 Marines, 500 Berber mercenaries, and 100 camels on a 500 mile march from Cairo, Egypt against the Tripolitan port city of Derne. Their mission was to oust Tripoli's ruling Pasha, Yusut Karamanli, who had the flagstaff from the U.S. Consulate cut down after taking power from his older brother, Hamet (a Pasha sympathetic to America). The battle on April 27, 1805 became the decisive action of the First Barbary War. Supported by naval gunfire from the USS *Argus* and *Hornet*, First Lieutenant Presley O'Bannon (the Commander of the Marines) fought with such intrepidity that Hamet gave him a Mameluke sword, which now serves as the pattern for Marine Officer swords.

- **Belleau Wood** - From June 1-26, 1918, the Battle of Belleau Wood raged near the Marne River in France, and it proved to be a turning point in World War I. The Great War had been fought for over four years, and the Allies were on their heels after the Germans launched a massive spring offensive in 1918. On June 6th at 3:45 a.m. the Allies took the initiative away from the Germans when the Marine Brigade captured Hill 142 in bitter fighting. By June 26th the Marines had almost completely cleared the German troops out of Bois de Belleau (Belleau Wood). This was the first contact between American Expeditionary Forces (AEF) and German troops. The Marines suffered 7,253 wounded, 9,063 casualties, and 1,062 battle deaths. As a sign of respect for the Marines' bravery, the French Army officially renamed Belleau Wood the "Bois de la Brigade de Marine." To this day Marines actively serving in the Fifth and Sixth regiments wear the French Fourragère on their left shoulder to honor their WWI predecessors. Sergeant Major John Quick received the Navy Cross and Distinguished Service Cross for his heroism at Belleau Wood (he previously earned the Medal of Honor in 1898 at Guantanamo Bay, Cuba during the Spanish-American War). The future Commandants who fought here were General Wendell Neville (14th), General Thomas Holcomb (17th), General Clifton Cates (19th), and General Lemuel Shepherd (20th).

- **Saipan** - After taking the Marshall Islands during World War II, the American high command sought to capture the critical Mariana Islands (Saipan, Tinian, and Guam). On June 15, 1944, barely a week after the Normandy D-Day invasion in Europe, Marines stormed the beaches of the largest island (Saipan) in order to gain a crucial air base to launch long-range bombers directly at Japan's mainland. By July 9th, the 2nd Marine Division, 4th Marine Division, and the Army's 27th Infantry Division (Commanded by General Holland M. "Howlin' Mad" Smith) had defeated the 43rd Division of the Imperial Japanese Army, commanded by Lieutenant General Yoshitsugu Saito. Lieutenant General Smith declared it "the decisive battle of the Pacific offensive" and General Saito wrote that "the fate of the Empire will be decided in this one action." The lessons learned from Saipan, such as improvements in close air support, directly led to success in the Philippines, Iwo Jima, and Okinawa. American B-29 Superfortress bombers from Saipan soon began regularly attacking Tokyo.